Leaping Learners Education, LLC

For more information and resources visit us at:

www.leapinglearnersed.com

Every attempt has been made to credit each photo. Please contact us if there has been an error and we will resolve the issue.

Photo Credits

Monkey graphics © dreamblack46/stock.adobe.com

Pg. 3-4 © max_776/stock.adobe.com; Pg. 5 © kungverylucky/stock.adobe.com; Pg. 6 © photographee2000/ stock.adobe.com, Pg.10 © kleberpicui/stock.adobe.com; Pg. 11-12 © Jacinto/stock.adobe.com; Pg. 13-14 © © kungverylucky/stock.adobe.com; Pg. 17 © hstiver/ stock.adobe.com; Pg. 18 © rockerchick1080/stock.adobe.com; Pg. 19-20 © garden/stock. adobe.com

All design by Sean Bulger
All other pictures by Sean Bulger or royalty free from Pixabay.com

ISBN
978-1-948569-17-0

Dear Parents and Guardians,

Thank you for purchasing a *Matt Learns About* series book! After teaching students from kindergarten to second grade for more than seven years, I became frustrated by the lack of engaging books my students could read independently. To help my students engage with nonfiction topics, my wife and I decided to write nonfiction books for children. We hope to inspire young children to learn about the natural world.

Here at Leaping Learners Education, LLC, we have three main goals:

1. Spark young readers' curiosity about the natural world
2. Develop critical independent reading skills at an early age
3. Develop reading comprehension skills before and after reading

We hope your child enjoys learning with Matt. If you or your children are interested in a topic we have not written about yet, send us an email with your topic, and maybe your book will be next!

Thank you,

Sean Bulger, Ed.M

www.leapinglearnersed.com

Reading Suggestions:

Before reading this book, encourage your children to do a "picture walk," where they skim through the book and look at the pictures to help them think about what they already know about the topic. Thinking about what they already know helps children get excited about learning more facts and begin reading with some confidence.

Preview any new vocabulary words with your child. Key vocabulary words are found on the last few pages of the book. Have your children use the new phrases in their own words to see if they understand the definition.

After previewing the book, encourage your children to read the book independently more than once. After they have read it, ask them specific questions related to the information in the book. Encourage them to go back and reread the relevant section in the book to retrieve the answer in case they forgot the facts.

Finally, see if your child can complete the reading comprehension exercises at the end of the book to strengthen their understanding of the topic!

This book is best for ages 6-8
but. . .
Please be mindful that reading levels are a guide and vary depending on a child's skills and needs.

Matt Learns About . . . Sloths

Written by Sean and Anicia Bulger

Table of Contents

Hi! My name is Matt.
I love to discover and
learn new things. In
this book, we will learn
about sloths. Let's go!

Introduction

Which rainforest animal is super

S

L

O

W?

A sloth!

Habitat

Sloths live in Costa Rica and the Amazon rainforest.

NORTH AMERICA
Costa Rica
Amazon
SOUTH AMERICA

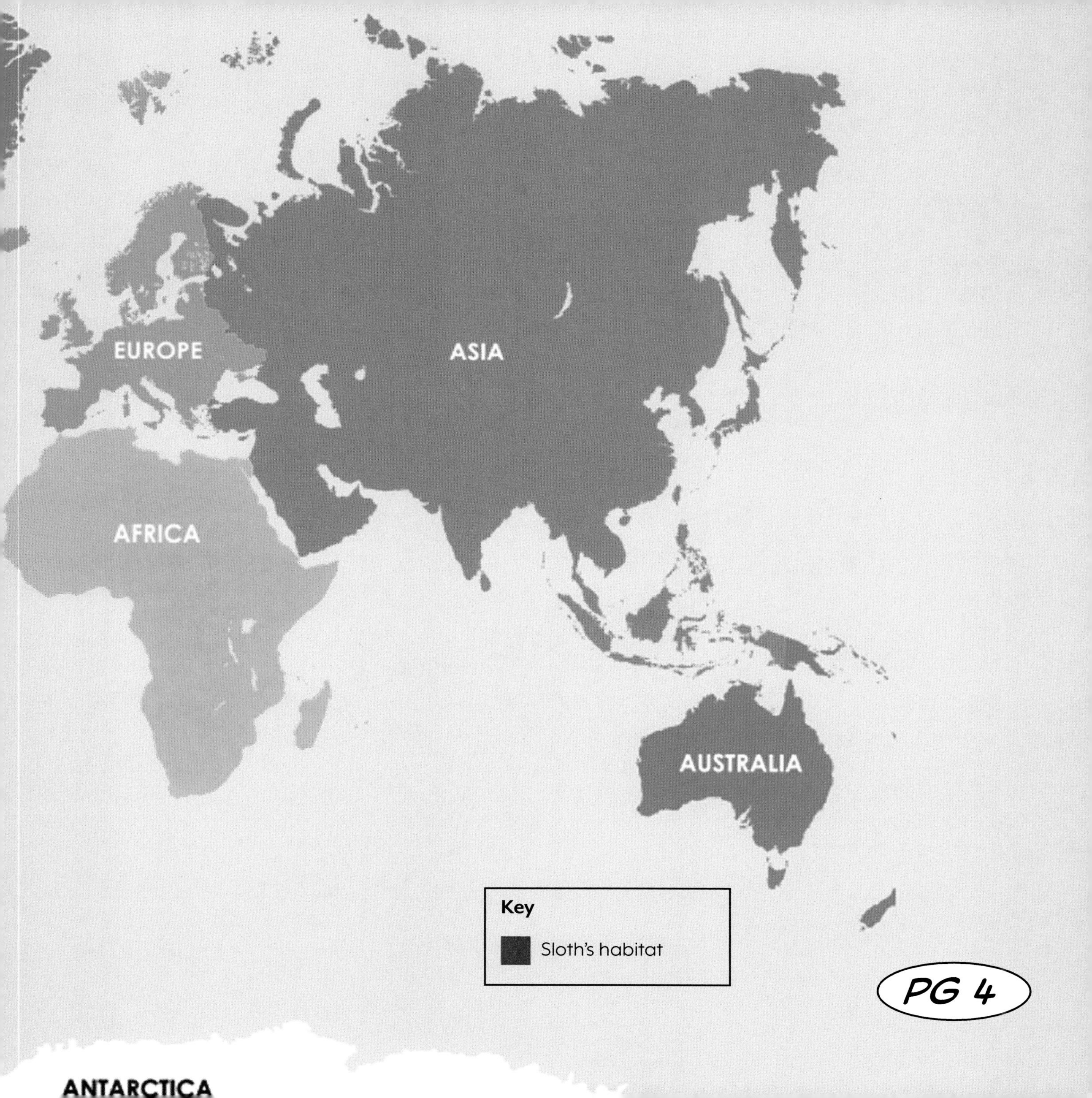
EUROPE
ASIA
AFRICA
AUSTRALIA
Key
Sloth's habitat
PG 4
ANTARCTICA

Where do sloths live?

Sloths live in trees in the canopy layer of rainforests. They hang from **branches**.

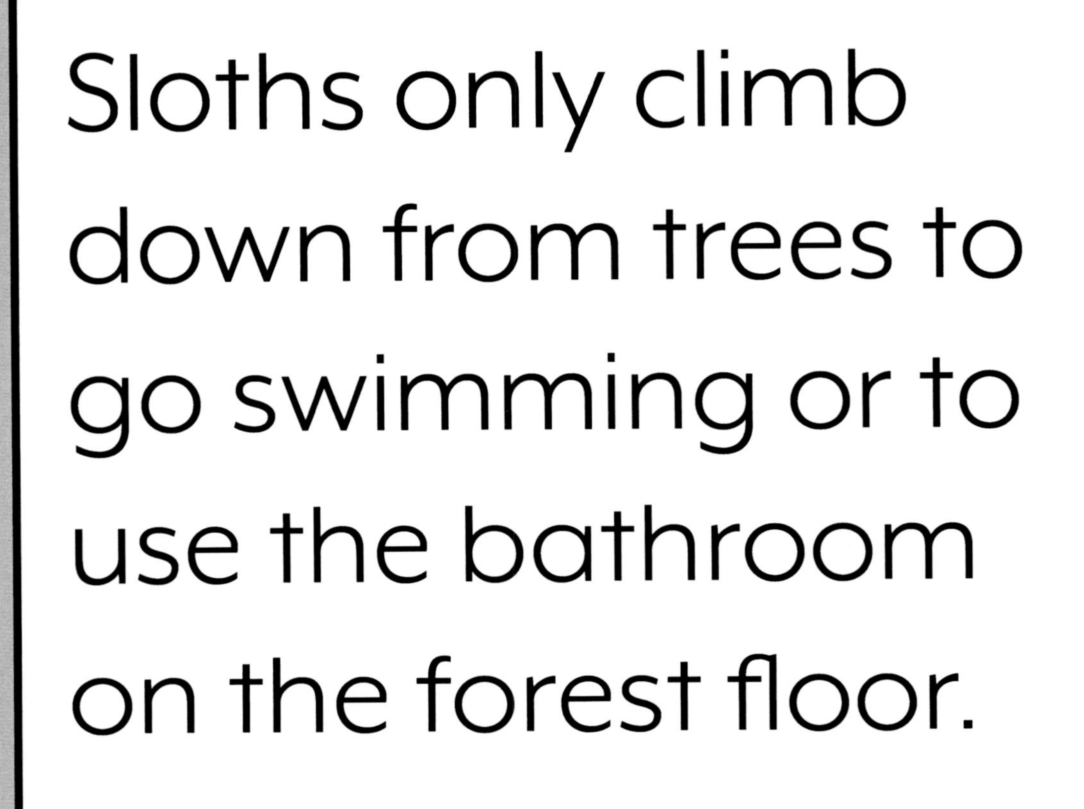

Sloths only climb down from trees to go swimming or to use the bathroom on the forest floor.

Body

Nose

Eyes

Sloths have two eyes, two ears, a nose, two arms, two legs, and long claws. Some sloths have three claws and some have two claws.

Sloths have gray or brown fur. Since sloths move slowly, some animals, like beetles and moths, live in their **fur**.

Beetle

Food

What do sloths eat?

Sloths eat leaves from rainforest trees. It takes a long time for sloths to **digest** their food because their bodies work very slow.

Predators

Some of a sloth's **predators** are big cats and large snakes. They hunt for sloths in the trees and on the ground.

Big cats
PG 16

Staying Safe

Since sloths move slowly, green **algae** grows on them.

This green algae helps sloths **camouflage** themselves with the green leaves and hide predators.

Sloths use their longs claws to grip branches and protect themselves when in danger.

Claws

Baby Sloths

Sloths have only one baby at a time. The baby stays with its mom after it is born. When it is older, it leaves its mom and lives alone.

Baby

Sloths are great swimmers.

Sloths only use the bathroom once a week.

A long time ago, sloths were the size of elephants!

Glossary

A glossary tells the reader the meaning of important words.

Branches – Thin parts of trees that grow leaves

Fur – Soft, thin hair over skin

Digest – Break down food in the stomach

Predator – Animal that eats other animals

Algae – Green plant-like organism that get their energy from the sun

Draw a picture of a sloth.

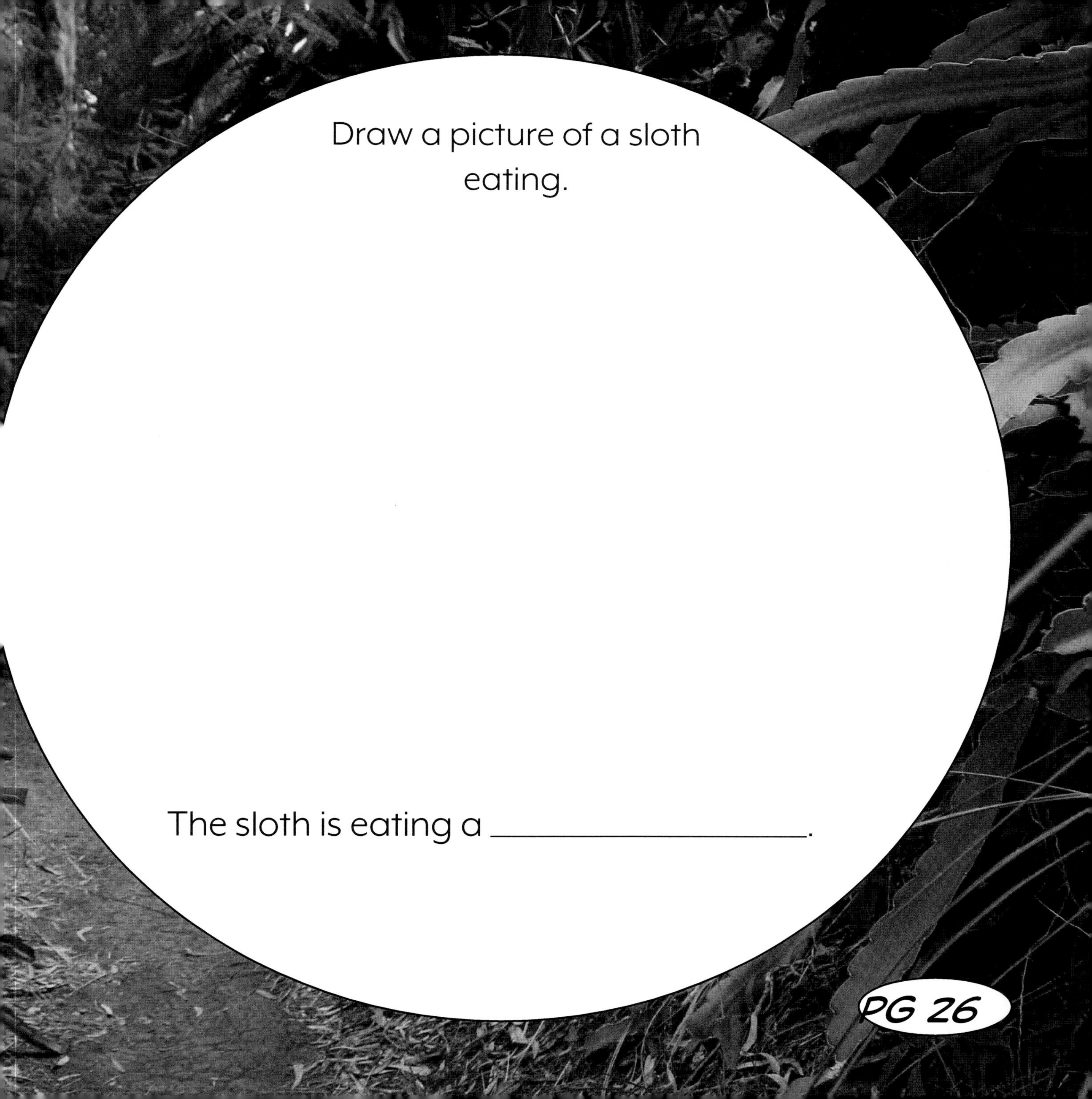

Draw a picture of a sloth eating.

The sloth is eating a ___________________.

Quiz

1. Which layer of the rainforest does the sloth live in?

a. Canopy

b. Emergent

c. Forest floor

2. What are two ways sloths use their claws?

a. To hang on branches, protect themselves

b. To carry food around, jumping

c. To hunt prey, climbing

3. What is the main idea of the section called "Habitat"?

a. What sloths look like

b. Where sloths live

c. What sloths eat

Common core standards:
RI. 1. 1 - Questions 1, 2
RI. 1. 2 - Question 3

4. How many babies does a sloth have at a time?

a. 1

b. 2

c. 3

5. How do algae help the sloth stay safe?

a. Algae smells bad

b. Algae is poisonous

c. Algae helps the sloth hide

6. What does the picture on page 12 teach you about?

a. Animals that live in a sloth's fur

b. Different species of sloths

c. How sloths climb trees

Common core standards:
RI. 1. 1 - Questions 4, 5
RI. 1.8 - Question 6

Want to learn about ocean animals? Check out the "Fay Learns About..." Series!

Great for emerging readers ages 6-8

Want to learn about Farm Animals? Check out the "Katie Teaches you About..." Series!

Great for early readers ages 4-6

Want to learn about colors? Check out the "Clayton Teaches you About..." series!

Great for early readers ages 4-6

Made in the USA
Coppell, TX
08 December 2021